STARTING OUT

BULLDOZERS

MEG GREVE

CREATIVE EDUCATION • CREATIVE PAPERBACKS

I0892455

CONT

ENTS

I See a Bulldozer 4

Rip Up the Road 6

Clean Up 8

A Strong Helper 10

Make a Noise 12

Bulldozer Words 14

Reading Corner 15

Index 16

I SEE A BULLDOZER.

It moves on tracks.

It crawls along
the dirt.

Bulldozers push and scrape.

The blade pushes dirt and rocks out of the way.

It makes a new road.

A bulldozer can use its ripper to break up an old road.

A bulldozer pushes away debris. It makes room for a new building.

Nothing stands in the way of a dozer!

The blade can knock down trees.

Powerful bulldozers change the land. They can help make a forest into a farm.

A bulldozer is stronger than a bull!

MAKE A NOISE

BEEP,

EEP, BEEP.

Can you make the sounds of a bulldozer?

Listen to these sounds:

https://www.youtube.com/watch?v=eFl2ga4hlFM

Now it is your turn!

BULLDOZER WORDS

debris: broken pieces left after something is destroyed

ripper: a metal piece on the back of a bulldozer that is used to scrape or break up hard surfaces

tracks: moving belts driven by wheels

READING CORNER

Baron, Amy. *The Gentle Bulldozer*. San Diego, Calif.: Yeehoo Press, 2022.

Leaf, Christina. *Machines with Power! Bulldozers*. Minneapolis, Minn.: Bellwether Media, 2021.

Mitton, Tony. *Amazing Machines: Big Bulldozers*. New York: Kingfisher, 2022.

INDEX

blade 4, 6, 10
building 8
dirt 5, 6
farm 11
push(es) 5, 6, 8
ripper 4, 7
road 6, 7
tracks 4

PUBLISHED BY CREATIVE EDUCATION AND CREATIVE PAPERBACKS
P.O. Box 227, Mankato, Minnesota 56002
Creative Education and Creative Paperbacks are imprints of The Creative Company
www.thecreativecompany.us

LIBRARY OF CONGRESS CATALOGING-IN-PUBLICATION DATA
Names: Greve, Meg, author.
Title: Bulldozers / by Meg Greve.
Description: Mankato, Minnesota : Creative Education and Creative Paperbacks, [2025] | Series: Starting out | Includes bibliographical references and index. | Audience: Ages 4-7 | Audience: Grades K-1 | Summary: "Bulldozers will introduce budding book learners to a noisy, colorful world with this new Starting Out title. Colorful photos, labeled diagrams, 'Make a Noise' section, glossary, and more ignite a passion for learning"-- Provided by publisher.
Identifiers: LCCN 2023059414 (print) | LCCN 2023059415 (ebook) | ISBN 9798889891673 (library binding) | ISBN 9781682775523 (paperback) | ISBN 9798889891796 (ebook)
Subjects: LCSH: Bulldozers--Juvenile literature. | Earthmoving machinery--Juvenile literature. | CYAC: Bulldozers. | Earthmoving machinery. | LCGFT: Instructional and educational works.
Classification: LCC TA725 .G75 2025 (print) | LCC TA725 (ebook) | DDC 629.225--dc23/eng/20240128
LC record available at https://lccn.loc.gov/2023059414
LC ebook record available at https://lccn.loc.gov/2023059415

COPYRIGHT © 2025 CREATIVE EDUCATION, CREATIVE PAPERBACKS
International copyright reserved in all countries. No part of this book may be reproduced in any form without written permission from the publisher.

DESIGN AND PRODUCTION
Design by Rhea Magaro
Production by Beeline Media and Design
Art direction by Tom Morgan
Printed in the United States of America

PHOTOGRAPHS by Shutterstock (Maksim Safaniuk, Dmitry Kalinovsky, Valentin Valkov, Juan Enrique del Barrio, asadykov, kuzmaphoto, Deyana Stefanova Robova, alberto clemares exposito, Ministr-84, Bohbeh, Benno Putro)